STONE

Resting stone
walled round
with slate
slight covering
of snow
melting

SCAUR GLEN,
DUMFRIESSHIRE
JANUARY 1993

STONE

Andy Goldsworthy

Harry N. Abrams, Inc., Publishers

Produced by Jill Hollis and Ian Cameron for Cameron Books, Moffat, Dumfriesshire, Scotland

Library of Congress Catalog Card Number: 94-70210

ISBN 0-8109-3847-2

Published in 1994 by Harry N. Abrams, Incorporated, New York

Printed and bound in Italy by Artegrafica, Verona

10 9 8 7

Harry N. Abrams, Inc.
100 Fifth Avenue
New York, N.Y. 10011
www.abramsbooks.com

Abrams is a subsidiary of

LA MARTINIÈRE
GROUPE

ENDPAPERS
Sea cairn
PORTH CEIRIAD, WALES
22 JULY 1993

PAGE 1
Slate arch – working photograph by Julian Calder
SCAUR GLEN, DUMFRIESSHIRE
26 MAY 1993

The text of this book has grown from lecture and sketchbook notes made over the last few years. Its writing has been like climbing a tree. There were times when I found it difficult to reach the next branch. I would like to thank Jill Hollis, Ian Cameron and Dr Terry Friedman for their help in the climb and for also respecting my wish that the climb should not be achieved by cutting off the higher branches. I cannot claim to have the reached the tree top but hope I have managed to get some way up and at least found a strong bough upon which to rest a while.

My thanks also to Julian Calder for taking photographs where my own fell short and to Phil Owen who has assisted me on many projects and installations.

Andy Goldsworthy is represented by: Galerie Lelong, New York; Michael Hue-Williams Fine Art, London; Galerie S65, Aalst, Belgium; Haines Gallery, San Francisco; Galerie Aline Vidal, Paris.

Acknowledgements

pp.7, 8-9, 50-51, 69, 71 'California Project', organised by Haines Gallery, San Francisco

pp.10-11, 40-41, 42, 43b, c, d, 52-53, 54-55, 56 'Mid Winter Muster', project commissioned by the Adelaide Festival and Adelaide Botanic Gardens, Australia

pp.26-33 commission by Laumeier Sculpture Park, St Louis, Missouri

pp.34-35 'Steel Cone', commissioned by Gateshead Council, Tyneside, with support from Northern Arts and assisted by Sculpture North

pp.38-39, 46-47, 108-109, 118-119 private commissions

pp.48, 56-57, 58-59 'Sand Leaves', project commissioned by the Chicago Arts Club

p.62 'Woodland', project organised by Galerie Lelong, New York

pp.64-65 'Black Water Stone', commissioned by Le Conseil Général de Seine St.-Denis

pp.70-71 'Two Autumns', organised by Tochigi Prefectural Museum of Fine Arts and Setagaya Museum, Japan

pp.72-73 'Hard Earth', organised by Turske Hue-Williams Gallery, London

pp.74-75 'Seven Holes', commissioned by Greenpeace UK, London

pp.76-77 'Wood Held Rock', project organised by Tochigi Prefectural Museum of Fine Art and the British Council, Japan

pp.82-83 'New Drawing', installation organised by Aline Vidal Gallery, Paris

pp.86-87 'Stone Sky', installation commissioned by Atelier St Anne, Brussels and the British Council, Belgium

p.101 project organised by Oriel Gallery, Cardiff, to be installed August 1994

pp.106-107 'Stone Gathering', commmissioned by Viscount Devonport with support from Northumberland County Council, Northern Arts and Natural Stone Products

pp.112-113 'Slate Dome Hole', organised by Royal Botanic Garden, Edinburgh

pp.114-115 'Two Folds', commissioned by Ile de Vassivière, Centre d'Art Contemporain de Vassivière

pp.116-117 'The Wall That Went for a Walk', commissioned by Grizedale Society Sculpture Project, Grizedale Forest.

CONTENTS

STONE

Fixed ideas prevent me from seeing clearly. My art makes me see again what is there, and in this respect I am also rediscovering the child within me. In the past I have felt uncomfortable when my work has been associated with children because of the implication that what I do is merely play. Since having children of my own, however, and seeing the intensity with which they discover through play, I have to acknowledge this in my work as well.

I had to forget my idea of nature and learn again that stone is hard and in so doing found that it is also soft. I tore leaves, broke stones, cut feathers . . . in order to go beyond appearances and touch on something of the essence. I would often start by clearing a space in which to work and put things – place was as closely cropped as the material.

I cannot disconnect materials as I used to. My strongest work now is so rooted in place that it cannot be separated from where it is made – the work is the place. Atmosphere and feeling now direct me more than the picking up of a leaf, stick, stone . . .

Inevitably materials and places gather associations and meanings as my work develops, but in ways that draw me deeper into nature rather than distracting me from it. What I could previously see only by working close up is now also visible to me from a distance. I now want to understand the untorn leaf, the unbroken stone, the uncut feather, the un-cleared space . . . and to perceive all materials as the same energy revealed differently.

I am no longer content simply to make objects; instead of placing works upon a stone, I am drawn to the stone itself. I want to explore the space within and around the stone through a touch that is a brief moment in its life. A long resting stone is not an object in the landscape but a deeply ingrained witness to time and a focus of energy for its surroundings.

My work does not lay claim to the stone and is soon shed like a fall of snow, becoming another layer in the many layers of rain, snow, leaves and animals that have made a stone rich in the place where it sits.

Although I occasionally work in wildernesses, it is the areas where people live and work that draw me most. I do not need to be the first or only person in a place. That no-one has gone before me would be a reason for me not to go there and I usually feel such places are best left.

I am drawn to wildness but do not have to be in a wilder-ness to find it. If much of my work appears to be made in such places it is because I find wildness in what is often considered commonplace. Going to other countries is interesting but not essential to my art. Most (if not all) that I need can be found within walking distance of my home. When travelling I regret the loss of a sense of change. I see differences not changes. Change is best experienced by staying in one place. I travel because I am invited and accept this just as I do ice when it's freezing and leaves in Autumn. The choice of where to work is never entirely within my control.

I am not a great traveller and when abroad I will settle into a daily routine of going back and forth to work in the same place. Even in the vast Australian outback I worked mainly in one area of a small hill. I returned there on my second visit and would be happy to go there for a third.

I revisit some stones, as I do places, many times over. Each work teaches me a new aspect of the stone's character. A stone is one and many stones at the same time – it changes from day to day, season to season.

I do not simply cover rocks. I need to understand the nature that is in all things. Stone is wood, water, earth, grass . . . I am interested in the binding of time in materials and places that reveals the stone in a flower and the flower in a stone.

It is difficult for a sculptor to work with petals, flowers and leaves because of their decorative associations. I cannot under-stand nature without knowing both the stone and the flower. I work with each as they are – powerful in their own ways – the flesh and bones of nature.

I feel the same about colour. Colour for me is not pretty or decorative – it is raw with energy. Nor does it rest on the surface. I explore the colour within and around a rock – colour is form and space. It does not lie passively or flat. At best it reaches deep into nature – drawing on the unseen – touching the living rock – revealing the energy inside.

Two granite boulders
one wrapped in manzanita branches
another clad with bark
lizards causing the bark to slip

LAKE TAHOE, NEVADA

22-24 MAY 1992

Sheep and kangaroo bones
bleached white by the sun
enclosing a boulder
snake

MOUNT VICTOR STATION, SOUTH AUSTRALIA

11 FEBRUARY 1992

Overnight frost
river ice
frozen to the shadow sides of a rock
becoming warmer
shedding the ice

SCAUR GLEN, DUMFRIESSHIRE
30 JANUARY 1992

Icicles frozen to rock – made over three days
I fell in on the first
found longer icicles on the second
heavy frost on the third

SCAUR WATER, DUMFRIESSHIRE

12-15 JANUARY 1991

Maple leaves held with clay
edging the stone, catching the sun

OUCHIYAMA-MURA, JAPAN

16 NOVEMBER 1991

Dandelions
on rock and riverside pool
where I laid beech leaves
the previous autumn

SCAUR WATER, DUMFRIESSHIRE

MAY 1993

Beech leaves

SCAUR WATER, DUMFRIESSHIRE

OCTOBER 1992

Dandelions

SCAUR WATER, DUMFRIESSHIRE

MAY 1993

Winter
ash

Spring
beech

Penpont stone

DUMFRIESSHIRE 1991-1993

Summer
sycamore

Autumn
hazel

Yellow elm leaves

laid over a rock

low water

SCAUR WATER, DUMFRIESSHIRE

15 OCTOBER 1991

Red maple leaves
held with water
sunny

OUCHIYAMA RIVER, JAPAN

18 NOVEMBER 1991

River rock

low water

only a small pool

to wet the rock and dip leaves in

before laying them around the rock end

LAUMEIER SCULPTURE PARK, ST LOUIS, MISSOURI

18 SEPTEMBER 1991

Same place as yesterday
river drained away
walled around the rock
returned three days later
raining but river still empty at first
suddenly flooding
like a tide coming in
as I started to dismantle the wall

LAUMEIER SCULPTURE PARK, ST LOUIS, MISSOURI
19, 22 SEPTEMBER 1991

Removed stones of previous work
wet leaves laid on wet rock
raining heavily
river rising
eventually lifting the leaves off

LAUMEIER SCULPTURE PARK, ST LOUIS, MISSOURI

22 SEPTEMBER 1991

Returned to rock in summer
worked round with dead sticks
river dry and quiet

LAUMEIER SCULPTURE PARK, ST LOUIS, MISSOURI

11 JUNE 1992

CAIRNS

I am reluctant to carve into and break off solid living rock, or to move a large boulder from the place where it has been for a long time, unless in a quarry.

I feel a difference between large, deep-rooted stones and the debris lying at the foot of a cliff, pebbles on a beach, stones rolled to the side of a field . . . These are loose and unsettled as if on a journey, and I can work with them in ways I couldn't with a long resting stone. To take such a stone would be like extracting a tooth and I would have missed the greater opportunity of knowing the stone in the place it has become part of.

I work in the quarries from which my house is built, with the tons of waste whinstone and sandstone that have been left there. I also use raw quarried stone indoors as a reminder of a building's origin and a lever to prise out its nature. Stone walls and slate roofs.

I am interested in the journey quarry stone has made: violently extracted from the ground, worked smooth by hand and machine, then laid carefully in a building.

I am aware of the skill and hard work of those who work in the quarries. My first slate works in the mountains of Cumbria

Steel cone

Photograph by Julian Calder

GATESHEAD, TYNESIDE

SUMMER 1991

and Wales have made the memory of the holes left there part of all the slate work I have done since. I acknowledge these connections by deliberately using quarry debris and waste building materials.

Construction-site roofing slate, rescued from a skip and worked into a cone, takes on some of the strength and power it had when bedrock – a reminder of where it began.

The journey has come almost full circle with recent (as yet unrealised) ideas to make cones in the underground caverns where slate was mined.

The scrap steel cone stands on the site of an old foundry (now gone) and touches the nature of an urban environment. Steel has a rawness that retains a quality of the earth from which it comes. I can feel its source. This cone draws strength

Cone to mark night becoming day
began around midnight
worked through until dawn
finished at daybreak

SCAUR GLEN, DUMFRIESSHIRE
24 OCTOBER 1991

and meaning from the nature of steel, city and a site that is now grown over and wooded where not so long ago people lived and worked. It links different natures and times.

Elsewhere there are partners to the steel cone made of stone, wood, ice, leaves and snow. Like the cairns that define paths in the mountains and fells of Britain, they are journey markers to my travels – leaving a trail. It is not necessary for all of them to survive, and some have been made to mark time or a quality of light, dusk, dawn, moonlight.

My first night works used the low temperatures to freeze ice to ice, ice to stone. They were completed by early morning and seen in daylight. I have now become interested in the night itself – work made at night, for the night and to be seen in the dark. Night totally changes my perception and has helped me 'see' more clearly during the day.

The cones are made more by feel than by calculation, especially those constructed at night. The resulting irregularities

give energy to form and help it fit into the place. They are made with the same shape in mind, but the response to each stone, place and time produces enormous variation. Repeating this work makes me more aware of the differences.

I am fascinated by the way a cone grows, stone upon stone, layer by layer – as a tree does, ring upon ring. By making slight changes in the placing of each stone, the shape can be brought out or taken in, made elegant or squat, full or empty.

I enjoy the unpredictability of working by eye and hand. Tools and machinery are used for the larger, more permanent sculptures, but for the most part my hands are the best tools I have. I need direct contact between hand and earth. There is enormous freedom in going empty-handed to a place and discovering there the material and the means to work with it.

Each time I try to achieve a perfect cone but somehow always lose control in the making. Cones dictate their own shape and I resist making 'corrections' which might interrupt the

Cone to mark day becoming night
began around mid morning
worked through until dusk
finished at the day's end

GLENLEITH FELL, DUMFRIESSHIRE
22 OCTOBER 1991

flow of form. Irregularities must be worked through slowly to avoid harsh or abrupt changes. I must think beyond the detail of individual stones to an overall idea of the form. The day I make a perfect cone will possibly be the last time I make one.

The cones are built solid for both practical and aesthetic reasons. The form is an expression of the fullness, vigour, heavy ripeness and power of nature generated from a centre deep inside – the seed becoming a tree and the unfolding of a flower. This feeling of endless layers of growth and internal depth would be lost if the cone were hollow. A concentration of energy is achieved when materials are drawn tightly together. It is this underlying energy that truly binds each piece to the next.

On occasion, I have reduced the cone to a simple pile so that it becomes more rooted in and dependent on place – less self-contained. These cairns are more passive and quiet, needing place to make them active. I have collected stones that glowed white under a full moon, pebbles at low tide,

earth-stained stones at the end of the day, and worked when it was cold enough to freeze one stone to another.

Each pile is much more than its material. There are cairns of light, colour, cold, water . . . and stone. I have gathered the moonlight, setting sun, incoming tide, hard frost . . .

One cairn next to a river in Illinois marks the '100-year flood' that happened in 1954, and is built to the height the water then reached. The twelve-foot-high cairn gives a feeling of the weight, power and volume of a river in flood in a way that a marked pole never could. From now on annual high-water levels will be carved into the cairn as horizontal lines. The rhythm and position of these lines will be dictated by the rhythm of the river, and their making could stretch well beyond my lifetime.

Four limestone corner cones
one leading to another
photographed at midday

Working photograph by Judith Goldsworthy
Cones on opposite page
photographed by Jean-Marc Pharisien

NICE, FRANCE
SPRING 1993

Difficult to work the round pebbles
first attempt collapsed
made just out of reach of the tide
remained for several weeks
until higher tides

KANGAROO ISLAND, SOUTH AUSTRALIA

26 FEBRUARY 1992

River cairn
raining heavily

SCAUR WATER, DUMFRIESSHIRE
2 AUGUST 1993

Cold wet river stones carried without leaving drips
made into a cairn on a dry pebble beach

SCAUR WATER, DUMFRIESSHIRE

1 SEPTEMBER 1993

Orange stones
gathered for the setting sun

MOUNT VICTOR STATION, SOUTH AUSTRALIA

23 JULY 1991

Cairn to follow colours in stones
for the day

MOUNT VICTOR STATION, SOUTH AUSTRALIA

23 JULY 1991

Stones gathered at night
for the moonlight

MOUNT VICTOR STATION, SOUTH AUSTRALIA

23 JULY 1991

Stones
dipped in water
then frozen to quarry face
made slowly over three cold days
place untouched by the sun in winter

PICKERING, YORKSHIRE
23-25 DECEMBER 1992

Field and gravel-pit boulders
piled up
to height of river flood in 1954
annual flood levels from now on
to be carved as lines along the cairn

Stonework by Joe Smith, assisted by Don Hausler
and Rick Rogers

FOX RIVER, ILLINOIS
NOVEMBER 1992

For the moonlight
boulders of sand
dug, scraped, piled, compacted

LAKE MICHIGAN, MICHIGAN

24 AUGUST 1991

SAND STONE

Dumfriesshire has a red sandstone which can be split along its grain into slices and is one of the few materials that I carve. I have cleaved quarry stones into layers, cut holes into each slab, then reconstructed the stone with the holes aligning, becoming smaller as they deepen . . . looking into the centre . . . back in time.

Even loose sand has some of the character of stone. Sand dunes form a compressed crust after rain which dries hard and is good for excavating. I have compacted sand into boulders, leaving them for the wind and water to work on.

I have never found a sand so tactile as in the red Australian bush. One morning after rain I began working the sand in my hand. Looking up, I saw a dead, grey, barkless mulga tree. I smeared sand onto the trunk, not really thinking it would stay, but discovered that it held in as if that was what it was intended for. I climbed the tree and started covering it from the top, jumping down carefully so that the tree didn't shake and throw off the sand. Intensely dark rain clouds gathered over the horizon against which the sun illuminated the tree. A heavy distant rain created a rainbow that arched over the tree. It was an extraordinary work – covered in an earth that hardened as it dried, becoming a stone tree in a very ancient land. At the time it all seemed so right and it is only when looking back that I am astonished that the moment and work happened at all – a meeting of tree, earth, rain and sun.

I go out every day I can. I make a lot of bad work with many failures. Impossible moments such as the red sand tree or a snake emerging from a bone boulder are somehow made inevitable by establishing a rhythm in my work that is driven by chance, intuition and instinct. These are beautiful moments that I live for, when my work goes to the heart of a place.

I strive for a beauty that cannot be explained in words. It is too hard-won to be described by conventional definitions. I am drawn to beauty as a tree is drawn to light or an animal to water. It is nourishment and a reality that does not ignore those qualities that are considered ugly but touches on truth in the nature of things.

In the inevitable struggle of making art, the line between working with rather than against nature is difficult to define. Sometimes I have to question what I am doing and break with familiar ideas and forms to make work that is a shock to my perception of nature.

Occasionally I must lose control and explore the unknown and random, alert to the possibility that the difficulty I have fought against is potentially more interesting than what I am trying to achieve. Chance events such as a falling stone stack, the picking up of a leaf by the wind, a rising river, a cyclist riding over a dry shadow and leaving wet tyre tracks, washing my hands in a pool after working with red earth and making red clouds . . . can lead to a stronger work. I have to think beyond my first reaction that the work has been spoilt.

Ideas are often revealed in a moment. Journeys are initiated in pursuit of a fleeting glimpse which then needs rediscovering – to draw out what I feel is there but don't understand. A good work is a moment of intense clarity, not mystery.

Seeing clearly in a chaotic situation is the means by which an artist becomes a participant and gains control. It is the difference between a ship sinking or sailing in rough seas. I cannot change the force, but I can be witness to it. The intention is not to tame the chaos but to tap its energy.

When first made, new work can appear disconnected from what I have done before, but it eventually becomes part of a natural progression, and links between different orders and structures are established. It is easy to make a mess. I want my work to be taut and I am not interested in making weak arrangements of nature in the pretence of being sensitive.

I do not accept all that I make. Even something as unpredictable as a throw creates its own demands that result in a good or bad work. I know that if thrown into the right time and space the work will be strong. Choosing the moment is as important as the throw itself.

Form should be at least as strong as the idea. Wet sand simply picked up and thrown makes lumps. By cupping the sand in the water and making a well of liquid sand from which it is lifted, then squeezing my hands together as it is thrown, fluid drawings are produced.

It took many throws in different seasons, time and light to achieve deep red splashes. The first were made in winter when the river reflected little growth on the trees. I waited for summer when the place became green, making the red more intense by its contrast. I waited for evening when the sun was low and shining on the trees and the splash.

The balls of earth are compressed, then held in the water before being thrown so that the splash is fluid. A splash is made slowly. The intensity of a moment affects the perception of it. In making a splash I have time to throw the earth, wash my hands, reach for my camera, let the earth hit the water, then wait for the secondary splash (the first is too white) when the colour becomes mixed with the river and erupts from below.

Sand stones

HOPE RANCH BEACH, CALIFORNIA

16 MAY 1992

My intention is not to improve on nature but to know it – not as a spectator but as a participant. I do not wish to mimic nature, but to draw on the energy that drives it so that it drives my work also. My art is unmistakably the work of a person – I would not want it otherwise – it celebrates my human nature and a need to be physically and spiritually bound to the earth.

If clarity of outline is a criterion for a good shadow, then sand is the best surface to lie down on. It accepts the body comfortably and soaks up the rain without allowing it to spread into the dry shadow.

Stone is much more difficult. The form of the rock distorts the figure, sharp points add to the discomforts of cold, itching and mosquitoes that have to be endured without moving over the often long time it takes for the surrounding ground to become wet.

With sand, I can lie for a long time and still leave a clear, dry shadow. On stone, timing is more critical. I have to lie down only for the time it takes for the stone to become wet and get up before the wet runs underneath my body. This usually means getting up while it is still raining which gives time only for a glimpse of a dry shadow before it fades to wet. Each stone is different: granite accepts less water than sandstone.

Sand is matt and less affected by light than stone. When lying on stone I have to consider quickly the light of a place. So many times a rushed decision has meant lying down, only to get up knowing my shadow is there but being unable to see it. Occasionally I have lost the shadow to the continuing rain before I can photograph it.

A place made dark by an overhanging tree or a building that reflects darkly on a wet surface reveals best the lighter dry shadow.

I never realised how unpredictable rain is. It is not often that I can choose where to lie down. I like the random manner in which place is chosen by the first raindrops.

I am not interested in wetting the rock other than by the rain. The purpose of the shadow is to know the rock and the rain. I feel the rain as the ground does. Each rainfall has its own

character and rhythm. Some rain has soaked me in seconds and given me only just enough time to lie down.

There are places and rocks where I would like to make shadows and when rain is imminent I work nearby. Sometimes I am unable to leave and remain bound to the rock by a tantalising spitting of rain. One day when rain was forecast I waited on a rock for several hours under the darkest of skies in the slightest of drizzles that dried up as soon as it fell. The rock was at an angle, and my muscles ached with the effort of gripping its surface. Eventually I gave up, knowing that as soon as I left (whenever that was) it would begin to rain heavily (which it did an hour or so later).

A perfect rainfall for sand comes hard and fast, wetting the ground thoroughly. For stone it is better for there to be a slow and gentle rain that wets the rock, but without causing the water to run, then stops and allows me the luxury of seeing the shadow in dry weather. This does happen, but not often.

So many times I have lain down and the rain has petered out too soon to leave a distinct shadow. I have begun to like these faint shadows which express light rainfalls as a strong shadow does heavy rain.

Although among the most passive of my works, rain-shadows are some of the most demanding and those that I have least control over. I have made winter shadows but see it becoming seasonal summer work when the rain is warmer and the stone dries more quickly. In winter a stone remains wet even during dry days. Summer days of sun and showers (with the stone drying quickly once the rain has stopped) allow several shadows in one day, sometimes on the same rock.

They are much more than an outline. I see all that I do as a shadow laid down in time as I lie on the ground. The trace that I leave will disappear from view, but is not lost – it becomes part of the place. This is also how I feel about life – the shadows we leave are evidence of our being here. They talk of nature in both city and country and are as appropriate to a New York sidewalk as to a mountain boulder.

Mulga branches
placed end to end
edged with sand
making a ridge to catch the setting sun

MOUNT VICTOR STATION, SOUTH AUSTRALIA

17 JULY 1991

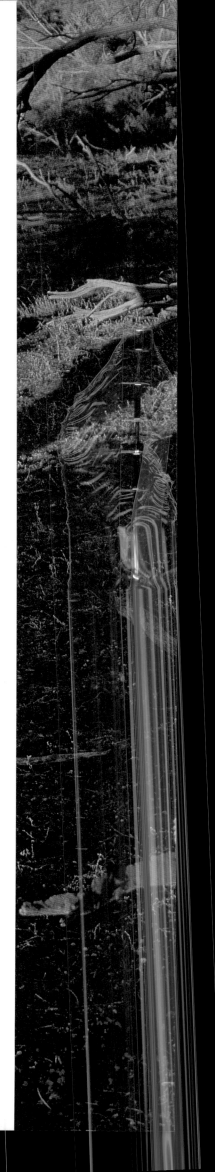

Rained overnight
damp sand
spread over the north side of a mulga tree
facing the midday sun

MOUNT VICTOR STATION, SOUTH AUSTRALIA

22 JULY 1991

Red sand

thrown into a blue sky

Photographs by Fiona Maclachlan

MOUNT VICTOR STATION, SOUTH AUSTRALIA

10 JULY 1991

Wet sand cupped, squeezed, thrown out of water

Photographs by Judith Goldsworthy

LAKE MICHIGAN, MICHIGAN

AUGUST 1991

Mud stones

red earth splashes

SCAUR WATER, DUMFRIESSHIRE

JULY 1992

Late afternoon – rain forecast – waited four hours – no rain

returned two days later – waited an hour and a half – began to rain gently

dust, stone becoming wet without running

laid down for half an hour

before getting up

CENTRAL PARK, NEW YORK

19 JUNE 1993

A stone
laid on previously
for several hours
slow rain dried
as quickly
as it wetted
returned in summer
sun and showers
five shadows
in three days

GATELAWBRIDGE,
DUMFRIESSHIRE

JULY 1993

TORN STONE

'And as it happened to a *Dutch* Ambassador, who entertaining the King of *Siam* with the particularities of *Holland*, which he was inquisitive after, amongst other things told him, that the water in his country, would sometimes, in cold weather, be so hard, that Men walked upon it, and it would bear an Elephant, if he were there. To which the King replied, "*Hitherto I have believed the strange things you have told me, because I look upon you as a sober fair man, but now I am sure you lye*".'

John Locke *An Essay Concerning Human Understanding* 1689

I feel a deep insecurity in nature – a fragile, unpredictable and violent energy. The black holes and cracks are windows into that energy.

Place has to be right for a black hole. The black is not achieved by the harsh light of a sunny day. It is a shadow within a shadow, black in black, and best made under a soft grey sky when everything is in shade.

Depth depends on the light to which the hole becomes an aperture – a bright place demands a deep cavity and a small opening. At its best, the relationship between space and light gives the appearance of the black rising up and buckling the rim under its pressure. In a building the black penetrates the floor, causing an eruption of surrounding earth – a reminder of the ground below and that nature is there too.

For many years I have hovered around the hole's edge – as if trying to find a way in. In 1992 I installed 'Hard Earth' in a London gallery using a white Dorset clay. The clay was smoothed out with a plasterer's float, and initially looked the same as the walls and ceiling: the gallery appeared to be empty. Gradually the nature of the clay revealed itself as it cracked, as if it was coming out of the building itself.

'Hard Earth' is the black released from its container. The hole held enormous potential for new discoveries. Its regularity of form begged to be questioned and broken.

What I find by chance and accident is often predicted in previous works by a progression that can perhaps be best explained through another recurring idea – the lines and cairns to follow colours in leaves or stone. Before starting it is difficult to predict what colours I will find or in which order they will be placed. Occasionally there is a missing colour – one or two stones between, say yellow and red. I know they are there – I just have to find them. I enjoy establishing an order that forces me to look hard to make it complete. When finished the colours flow and weld one stone to another. The same can be said of ideas. The aim is to understand the nature of nature – not isolated materials.

The direction the black has taken was hinted at in early works and anticipated in my writing.

In 1984, I wrote: 'Looking into a deep hole unnerves me and I am made aware of the potent energies within the earth. The black is that energy made visible.'

Five years later, I added: 'The black hole is like the flame of a fire. The flame makes the energy of fire visible. The black is the earth's flame – its energy. I used to say I will make no more holes. Now I know I will always make them. I am drawn to them with the same urge I have to look over a cliff edge. It is possible that the last work I make will be a hole.'

And in a diary kept whilst in the Arctic the same year: 'I keep referring to the place as a "landscape" – the landscape in which I work. And to some degree there are hills, but I am talking about water. When I was riding over the ice I thought, this isn't land, this is water, so is it a waterscape? But it is as solid as the earth, for the moment. The more I work with the snow and ice, the more I realise there is so much to learn

Black water stone

PARC DE LA COURNEUVE, FRANCE

SPRING 1993

about the land and and the processes and forces that make up the land and life. In that everything is fluid, even the land, it just flows at a very slow rate.'

I wrote about the black being like a flame and that the land is fluid before discovering by accident that stones gently fired in a kiln will slowly tear open revealing deep red cracks which turn black as the stone cools. It felt as if a key stone had fallen into place.

There is a violence in this work, but it comes from within the stones themselves. They have been returned to their origin and worked with the force that brought them into being.

It has taken many years for me to be able to work with fire, and I still feel cautious in its use. Fire looks good; I am less interested in the drama of the flames than in the fire's heart – the slow intense powerful heat that is at the core of nature.

The fired stones are not broken – they slowly tear open as they grow hot and start to melt. That stone can become liquid and liquid can become stone is a deeply unnerving but beautiful expression of change.

It is somehow appropriate that the extreme heat of a firing and the intense cold of the Arctic has made me aware of the flow of nature. It confirms my belief that I cannot understand cold without heat, water without stone, winter without summer . . .

In making the torn stones I feel I have entered the black which no longer needs to be defined by the hole. I have poured out the black but without it weakening or draining away. I am aware of the black in all things. By black I do not mean a pigment, but a depth and space in nature that flows through earth, stone, tree . . . the energy that drives change, and life. I describe this energy as black, but it is expressed in all colours and forms.

Cleanly broken pebbles
scratched white

SCAUR WATER, DUMFRIESSHIRE

APRIL 1987

THIS PAGE & OVERLEAF

Torn stones
Morecambe Bay pebbles

Fired by Barry Gregson

LUNESDALE POTTERY, LANCASHIRE

SUMMER 1990

Clay-covered rocks
cracking as clay dried
over several days
revealing the rock within

HAINES GALLERY, SAN FRANCISCO

JUNE 1992

Mashiko clay-covered
river stones

Photograph by Norihiro Ueno

TOCHIGI PREFECTURAL MUSEUM OF FINE ARTS, JAPAN

OCTOBER 1993

Hard earth – Dorset clay smoothed out, left to dry

Assisted by plasterer Brian Dick

TURSKE HUE-WILLIAMS GALLERY, LONDON

NOVEMBER 1992

Seven holes
Canonbury clay
dug from outside the back door
several months in drying

GREENPEACE UK OFFICE, LONDON

APRIL 1991

Granite boulder found split open

heavy rain

laid red leaves along the broken edges

returned a few days later

boulder dry and white

worked dark earth into innersides

heavily overcast

about to rain

going dark

MOCHIGASE-GAWA, JAPAN

30 OCTOBER & 3 NOVEMBER 1990

OVERLEAF

Peat rubbed into rocks
cold, raw fingers

HEALABHIAL BHEAG, ISLE OF SKYE

7 OCTOBER 1990

Peat

worked into

the underside of a mossy woodland stone

GATELAWBRIDGE, DUMFRIESSHIRE

5 APRIL 1991

Soft red stone
rubbed into the sides
of a seaweed-covered rock
between tides

HEYSHAM HEAD, LANCASHIRE

5 MARCH 1991

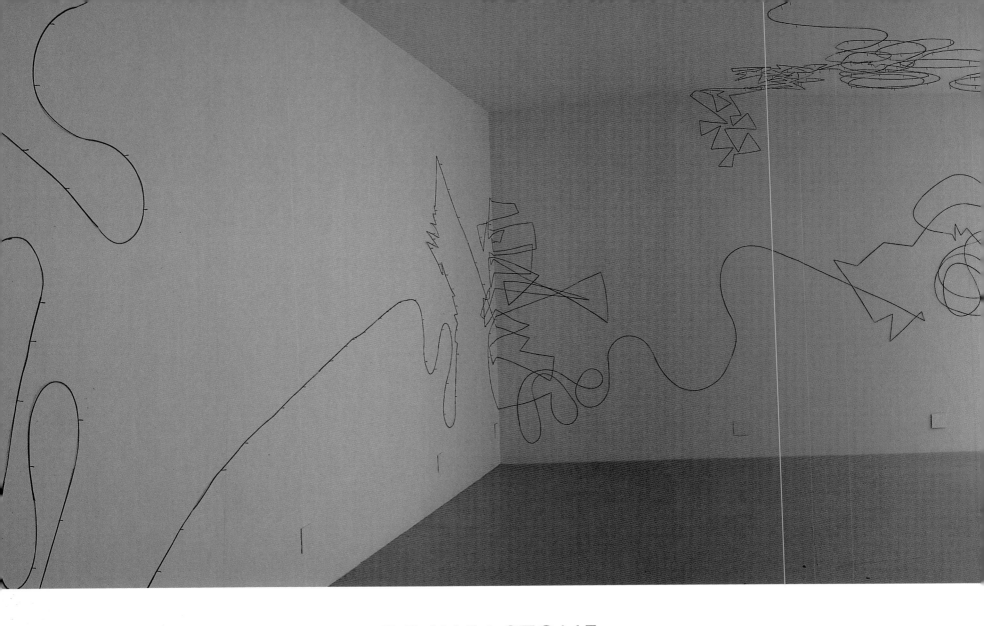

DRAWN STONE

Quick diary sketches are made when each work is finished. I also do proposal drawings. These are more concerned with information than with the activity of drawing itself, to which I feel there is a much deeper quality.

Drawing is not restricted to or defined by pencil and paper; it is related to life, like drawing breath or a tree taking nourishment through its roots to draw with its branches the space in which it grows. A river draws the valley and the salmon the river.

My days are defined by my work which leaves a trail that marks out my life. A leaf line is made knowing which direction I wish to take, but not always looking where I am going. The resulting line, with all its irregularities and diversions, describes the journey and place.

Drawing at its most essential is an exploring line alert to changes of rhythm and feelings of surface and space, playing from one to the other with the aesthetic rooted in movement.

A stone charges a place with its presence, with time filling in and flowing around it, just as a sea or river rock affects the surrounding water by creating waves, pools and currents. Drawing touches, works, explores and makes visible the relationship between rock and place.

My best drawings have been done with grass stalks, a stick on sand, scratched stones, snowball trails . . . lines that draw themselves as well as place. Stalks and scratches have qualities that are also acknowledged in the drawing. Their resistance gives vigour and tension to the line. A stalk can bend only so far before fracturing. All grasses are not the same, and I have learned to collect them from places where they grow sturdy and pliable.

By pushing the thin end of a stalk up the wider hollow end of another, I am able to make long and continuous lines. Early

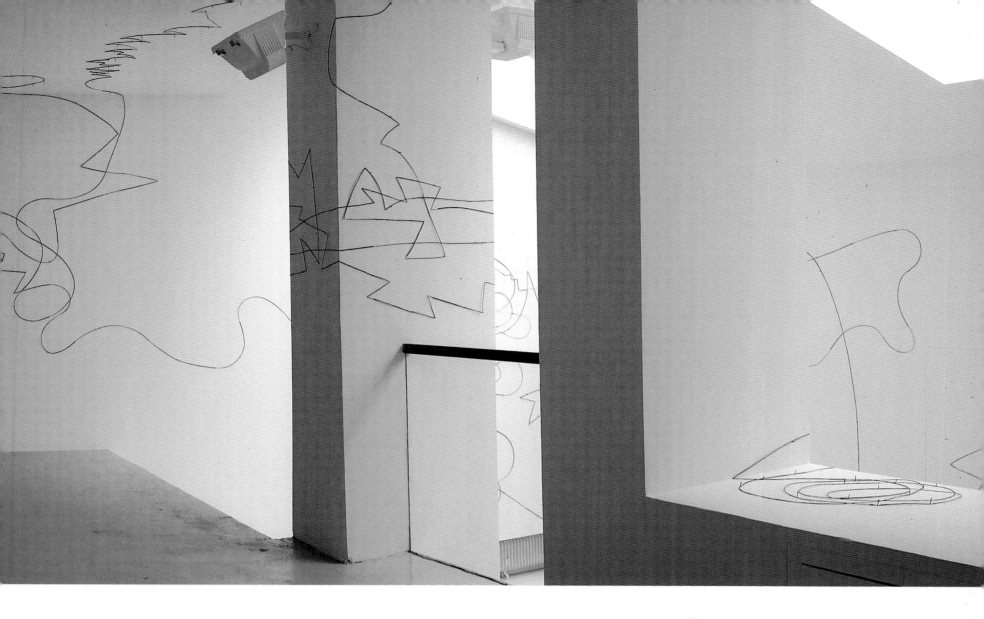

Rush line drawing pinned with thorns

ALINE VIDAL GALLERY, PARIS

NOVEMBER 1990

drawings were made against trees or the ground to which the lines were pinned with thorns. The tree was more a support than an integral part of the drawing itself. I had to learn first how to make a line before knowing how to use it.

Nine years ago I placed four stones on a path and covered each stone with mud to which I was able to secure grass stalks, so releasing the line into the surrounding space. It was, however, a space that I had made and the greater challenge lay in responding to existing rocks and spaces.

The increase in scale dictated by larger immovable rocks placed new demands on the drawing. Some works looked as if the line was tying the rock up, which was the opposite of my intention. I wanted a line generated by the rock – taut, alive and effortlessly flowing around it, not heavy or attached, let alone binding. The line must have its focus deep inside the stone. By working on its outside, I have become more aware of its internal space.

I have sometimes drawn lines and colour out of stone by scratching one against another, stone on stone, grinding out a dust, turning the hard rock into something closer to air or clouds.

To make the line or colour in any other way would forfeit its deeper significance. It is the underlying flow of colour and form that I want to understand: colour ground in a river pool which has itself been hollowed out by the grinding of stone. There is a shock in seeing a river rock pool turned red by the rubbing of two stones together. If to shock were my sole purpose, then it would not have mattered how the colour had been made. It is important to me that beyond this initial shock is the truth that it is a colour of the river in a landscape stained with a red that becomes more intense as I approach its source.

SWINDALE BECK WOOD, CUMBRIA

10 DECEMBER 1984

Rushes and grass stalks
thin end of one pushed up wider hollow end of another
lines secured with thorns and clay
drawing rocks on calm days

SCAUR GLEN, DUMFRIESSHIRE

25-26 JANUARY 1991

SCAUR GLEN, DUMFRIESSHIRE

24 JANUARY 1991

SCAUR WATER, DUMFRIESSHIRE

30 SEPTEMBER 1990

SCAUR GLEN, DUMFRIESSHIRE

3 JANUARY 1991

Stone sky
slate scratched on slate
drawing white

ATELIER ST ANNE, BRUSSELS

SPRING 1992

River rock
drawn over with a soft red stone
rubbed and wetted
lowered into the water
gave off surface cloud of red
gradually cleared

SCAUR WATER, DUMFRIESSHIRE

JULY 1992

Red river rock pools
soft red stones
rubbed together in water
staining each pool

Working photograph by Julian Calder

SCAUR WATER, DUMFRIESSHIRE

SPRING, SUMMER 1993

Two Scaur Water snowballs
stained red with Morecambe Bay stone

PENPONT, DUMFRIESSHIRE
WINTER – SUMMER 1992

HANGING STONE

'Here the stone leapt up from the plain earth, leapt up in a manifold, clustered desire each time, up, away from the horizontal earth, through twilight and dusk and the whole range of desire, through the swerving, the declination, ah, to the ecstasy, the touch, to the meeting and the consummation, the meeting, the clasp, the close embrace, the neutrality, the perfect, swooning consummation, the timeless ecstasy. There his soul remained, at the apex of the arch, clinched in the timeless ecstasy, consummated.

'And there was no time nor life nor death, but only this, this timeless consummation, where the thrust from earth met the thrust from earth and the arch was locked on the keystone of ecstasy. This was all, this was everything. Till he came to himself together, in transit, every jet of him strained and leaped, leaped clear into the darkness above, to the fecundity and the unique mystery, to the touch, the clasp, the consummation, the climax of eternity, the apex of the arch.'

D.H. Lawrence *The Rainbow*

When I was a student working at Morecambe Bay, Lancashire, I kept returning to Heysham Head and in particular to one rock. It stood waist high with a flat top which I used as a workbench.

Heysham Head looks out across the bay towards the Lake District, and I used the distant view of the mountains as bait to draw the eye through my work which often had holes as windows, into the space beyond. I became conscious of the negative shape the work cut into the sky and the space between and around the stones.

I would start early in the day when there was often little or no wind. This encouraged me to make delicately balanced work that explored the surface tension of a calm day. I got to know intimately all the ridges and cracks of the workbench stone that could provide a grip for stones stood on end. It was like a landscape: the stones I placed there took on monumental qualities and, although relatively small, touched on larger feelings towards nature. I would gently grind a stone until it found a hold. Letting go and finding that it stayed was a great moment. Less so if it fell.

I was deeply shocked to return a year or so later to find the workbench stone smashed by another that had fallen from the cliff above. It was a good lesson. Even though my work explores precariousness in nature, it never occurred to me that the very stone upon which I worked was soon to be shattered.

I can understand why the rock fell. What is more profoundly challenging and disturbing is that until it fell it was as if it never would.

I am reminded of a road running through a valley at Ashio, Japan, where I was working. A huge stone suddenly fell off the mountains opposite, bounced on to the road, then continued to the river below. A few minutes later a car passed along

Slate arch
second attempt
moving gently in the wind
too dangerous to leave standing
removed wedging stones
causing it to fall

Photographs by Julian Calder

SCAUR GLEN, DUMFRIESSHIRE
26 MAY 1993

the road, its occupants unaware of the drama that had just taken place.

Nature contains so much pent-up energy waiting to be released. Most boulders are either on or resting from a journey. I enjoy visiting stones marked on maps as 'Hanging', which appear ready to fall but can remain like this for hundreds of years – balanced between an instant and timelessness.

I am drawn to the tension that accumulates at the tip of a rock or the peak of a mountain. A rock balanced on a point feels supported by that tension.

A good work releases a build-up of energy with only the slightest of touches. It is sometimes less about how much I touch than where – like plants with seedheads that become taut as they ripen and need only a stroke to make them explode.

I experience the vigour and force of stone in the arches that I make, one side clasping the other in a grip that is almost equal so that neither gives way. Gradually, as in arm wrestling, one eventually weakens and both collapse.

I like the imbalance and tension of the unequal arch, and I do not usually intend it to last for long. These arches are as two stacked stone columns falling into each other and are as much about movement as architecture.

The sides are built over a supporting pile of stones which can be removed once the two meet. The arch becomes alive as it settles and takes its own weight. At best I wedge the arch so that it braces away from the supporting stones, allowing me to take them away easily. Sometimes I have to bodily hold, push, pull and twist it back into shape as an osteopath would a person.

If an arch does not collapse of its own accord, I sometimes weaken it so that it falls. I can learn as much by its destruction as from its making.

I have been surprised by how elastic an arch is. Even those which move in the wind show enormous tenacity before falling. Movement is their strength.

Out of the quarry
seven arches
made over two days
no failures
one almost fell –
slipped down the face of a rock
as I removed supporting stones

Assisted by Barry Gregson

GATELAWBRIDGE, DUMFRIESSHIRE

13-14 FEBRUARY 1993

Over the stone

Assisted by Wallace Gibson

SCAUR GLEN, DUMFRIESSHIRE

18 JANUARY 1993

Out of the stone

SCAUR GLEN, DUMFRIESSHIRE

17 JANUARY 1993

Between two trees

DUNBAR, PENNSYLVANIA

28 NOVEMBER 1992

Over the wall – five attempts over three days

SCAUR GLEN, DUMFRIESSHIRE

1 FEBRUARY 1993

Tree arch – river stones

SCAUR WATER, DUMFRIESSHIRE

FEBRUARY 1993

Balanced rocks
brought down by the incoming tide
those in a line
bouncing and banging
as they fell
then rolled around by the sea

TALISKER BAY, ISLE OF SKYE
11 OCTOBER 1990

PORTH CEIRIAD, WALES
23 JULY 1993

Intended to collect stones for 'Stone gathering'
but trucks and crane unable to make last few yards
to the quarry because of snow
stayed after they had gone
calm
balanced rocks

HIGH NICK QUARRY, NORTHUMBERLAND

4 MARCH 1993

Early morning calm
balanced stone
soon fell

HEYSHAM HEAD, LANCASHIRE

NOVEMBER 1978

WALLS

The British landscape is rich with agricultural dry-stone walls. Their effect upon my work has been more profound than that of prehistoric stone circles, to which I make only occasional visits. Walls are a living part of the landscape and are a great lesson to a sculptor in their response to material and place.

I have now made eight walls – all with Joe Smith and occasionally with Philip Dolphin, two of the best wallers in Britain. I used to repair walls when I worked on farms, and when making the first wall thought Joe and I might work together. However Joe kept taking my stones off and was right to do so. Good walling is not done casually and it has now become important to me that I don't make the walls. When I work with Joe I work not only with his years of experience but also with the tradition of walling of which he is part. If I made the wall it would become something else (apart from a less well-made wall). I want a simple (but beautiful) wall.

I help with the digging and labouring. It is for me to put the walls where they will take on new meanings that articulate a changing relationship between people and place. These meanings are made stronger because the walls are agricultural. Although my walls are not entirely bound by the demands of agriculture it was important that my first was made as a practical response to the need to divide a piece of land. I must keep the walls' roots firmly in agriculture, not art, otherwise they could lose their meaning as works of art. I enjoy the aesthetic of the practical.

My second wall reworks a derelict section in the network of walls that run through Grizedale Forest in Cumbria. The old walls are evidence of when Grizedale was fields. Mine responds to the place as it is now and for about 150 yards weaves through the trees (it has become known as 'the wall that went for a walk'). It is alert to the lie of the land, taking a route that incorporates a rock or tree into its length rather than flattening the ground or cutting the tree down – a line in sympathy with the landscape through which it travels, drawing a place.

The enclosures that are part of my first wall have their origin in those used by hill farmers to gather sheep. I enjoy the feeling upon entering them – sometimes for shelter on the open and windswept fell. The space is made quiet and intense by the containing wall, giving a sense of protection and care. I am less interested in the formal contrast between the circle and surrounding landscape than in these connections to farming.

Although the walls have strong links with the past they are not backward looking. I live in a region that has many good wallers. Making a wall for me does not carry the nostalgia that it might for a French artist in an area where little dry-stone walling is being done (like Vassivière).

At Vassivière I worked on the remains of a wall that once defined a field but now, since the damming of the valley, runs from the wood into a lake. I have explored the boundary between lake and wood with a wall that is in itself a boundary. The work encloses and contrasts the two spaces. The differences in space, feeling and light within each enclosure underscore the changes that have taken place.

A wall in Britain becomes part of a network that nourishes and maintains the wall, without which there is no need to keep it in order. The Vassivière wall will collapse and in its turn become evidence of an island, just as the previous wall spoke of a valley and fields. For me the most powerful comments are those that state and see things as they are without trying to contrive a point. The wall evokes feelings for the eight villages that were drowned without purposely taking issue with the damming but acknowledges that event as necessary to understanding the nature of Vassivière. The past roots a work and makes it resonate with place. The people, trees, rocks and earth are its past, present and future – touchstones to the life, energy and nature of a place.

I have not previously worked in a place that has been as suddenly altered as Vassivière: the shock of what happened there can still be felt, a place in which nature has been so changed yet where it is still strong. The wall is made on an

Stone gathering

Stonework by Joe Smith and Philip Dolphin
Photograph by Julian Calder

RAY DEMESNE, NORTHUMBERLAND
SPRING 1993

edge laid bare by the meeting of two natures – wood and water – and by the fluctuations caused by the dam, weather, and season. Water will erode the wall on the shore, and tree roots will undermine the woodland enclosure. The tension between lake and land (what was before and how it is now) will also work on the wall.

The wall should not fight these changes, nor is it designed to collapse. It has been made strong and if it stood in a field would last for generations. Changes that occur will be real, not effected by me through a deliberately fragile wall.

Its absence when collapsed will be more potent because of its strength when standing. Although I cannot predict what will happen, any changes will become part of the work. Even when utterly collapsed and evident only as a line of rubble, it will remain complete and finished.

It would be wrong not to acknowledge animals in a work that owes its origin to stock farming. One enclosure contains randomly placed stones – as animals in a confined space. Although not intended to represent sheep, the work draws together the energy of both stone and animal – an avalanche of birds, a flock of rocks, a herd of arches . . . The impact of animals upon my work, however, is not confined to their relationship to agriculture. Many a dull day's work has been invigorated by the sight of a deer, snake, buzzard, salmon, fox or seal. Animals are as much a force in nature as the weather, rivers, sea – another expression of nature's vigour and movement . . . moulding, digging, chewing, marking . . . working and changing the land.

Rock fold, excavated bed rock

Stonework by Joe Smith   Assisted by students from Carlisle College of Art

BARFIL, DUMFRIESSHIRE

WINTER 1993

Slate dome hole

Wall by Joe Smith

Photograph by Catriona Grant

ROYAL BOTANIC GARDENS, EDINBURGH

SUMMER 1990

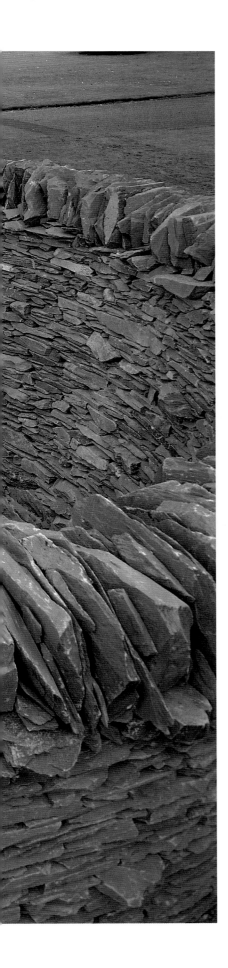

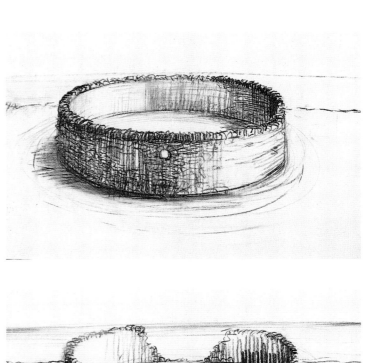

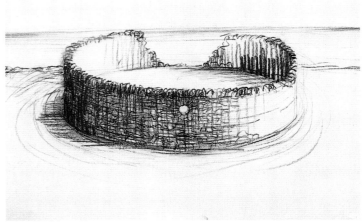

Proposal (unrealised)
for cliff top enclosure
to collapse a little each year
as the cliff eroded

SCARBOROUGH BLUFFS, CANADA

1992

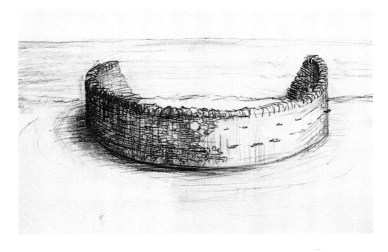

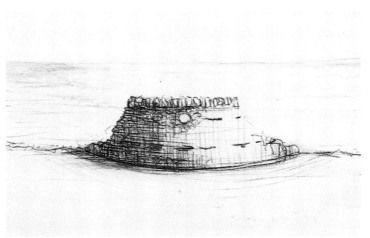

Two folds

Stonework by Joe Smith and Philip Dolphin   Photographs by Jacques Hoepffner

ILE DE VASSIVIÈRE, FRANCE

SPRING 1992

'Wall that went for a walk'

Stonework by Joe Smith, Philip Dolphin,
William Millar, Brian Metcalf

GRIZEDALE, CUMBRIA

SEPTEMBER 1990

Room

Stonework by Joe Smith

DUNBAR, PENNSYLVANIA

WINTER 1992

THE PHOTOGRAPH

I have a social and intellectual need to make photographs. As Brancusi said, 'Why talk about my sculpture when I can photograph it?'

Photography is my way of talking, writing and thinking about my art. It makes me aware of connections and developments that might not otherwise have been apparent. It is the visual evidence which runs through my art as a whole and gives me a broader, more distant view of what I am doing. On the occasions when film has not come out, that work feels dislocated – like a half-forgotten memory. The rigorous test which the photograph gives my sculpture confirms its success or failure. To interpret the relationship between the work outside and its image by deciding which is the art is too simple.

More practically (and possibly more significantly) the photograph provides a necessary barrier between the making and public viewing. It is a question of intention. I am not a performer. I usually go out by myself or with people I know. I need the freedom to work where and when I want, to change direction and to concentrate as I can only when alone. I work in public places (sometimes very public), and people often come across me by chance. This is a natural part of being outside. To make this my intention, however, would undermine the reasons for doing my art.

A public dialogue can be creative but I feel this is better explored through gallery installations and what I call my 'permanent work'. Ephemeral work made outside, for and about a day, lies at the core of my art and its making must be kept private.

I construct the image after the work's completion. During the making I become aware of the relationship with the surroundings . . . a nearby tree, rock, mountain . . . which needs to be explained in the photograph. Sometimes it is a particular movement, light or moment with which the work aligns that is important. I have laid works in wait to be activated by time and light: a rising sun, an incoming tide, a drop in the wind . . .

These elements often determine how and when the photograph is taken.

It is difficult to draw back from the rich textural quality that is so strong in my mind after working close up, and usually each work is represented by two views. One photograph, which will be accompanied by text, is taken from a distance and the other shows the work from much closer.

That one is shown larger than the other does not imply that it is more important. I feel that the visual smell and detail of the work is better expressed large.

It is appropriate for me to use a medium that is connected to time. A photograph roots itself in the moment when it was taken and in this respect functions differently to painting or drawing. The photograph is time. If I had to describe my work in one word, that word would be time.

Photography has made us more aware of the passing of time. Images that show us growing older will be further dated by the patina of a changing technology that will make today's colours look crude and grainy just as photographs of the past appear to us now. This is how it should be.

The photograph does not replace but comes out of the working process and can be as much part of an artist's vocabulary as recorded sound is of a musician's.

The photograph is incomplete. The viewer is drawn into the space between image and work. A bridge needs to be made between the two. It is necessary to know what it is like to get wet, feel a cold wind, touch a leaf, throw stones, compress snow, suck icicles . . . often reaching back into childhood to when those experiences were more alive.

If the photograph were to become so real that it overpowered and replaced the work outside, then it would have no purpose or meaning in my art.

Rapidly thawing icicles
stuck into the leeward side of a wall

SCAUR GLEN, DUMFRIESSHIRE
31 DECEMBER 1992